On The Nature Of Space, Time, And Evolution

By

Jan Frederick Andrus, Professor Emeritus, *University of New Orleans, LA*

Table of Contents

In our model, if acceleration is zero, motion is straight-line. In the general theory of relativity, it is *assumed* that objects move along geodesics.

In our formulation, how does one determine which gravitational acceleration derives from which body? This will be discussed later.

Alternatively, one may employ luminosity and brightness in order to estimate distance. See Morrison (1963) for definitions. These would be very approximate, but they would be sufficient to help living things avoid danger, seek food, etc. According to the anthropic principle, life can survive only in an accommodating universe.

Events and Processes

We adopt an event ontology in which events are real things. Processes consist of events ordered in a logical manner employing continuity of quantifiable features of events. See Andrus (1987). Andrus' proposal is similar to that of John Locke, who taught that personal identity is a matter of psychological continuity, with memory being an important component. Events in the life of an animal consist of mental experiences, some of which may be subconscious. If an animal has a split-brain or multiple identities, then each of these identities may form its own process. There are also processes consisting of the life histories of molecules and other existent particles.

It is suggested by Andrus that the events of a process are instantaneous or *point* events: i.e., an animal is essentially different in every event of its life. Perhaps even so-called lifeless things have at least a scintilla of consciousness. This is a view of many heavyweights of philosophy. It is known as panpsychism. After all, our only contact with other entities is through conscious experience.

Every living thing has its own series of events. There is nothing that we know of to ensure that these beings experience an objective reality. In fact, recently, scientists have produced what they claim is a non-objective experimental result. In the Andrus formulation, cause and effect don't exist because the events in their processes are completely defined in our own continuity model.

Measuring Gravitational Acceleration

Measurement of gravitational acceleration requires the use of accelerometers and gravimeters. Accelerometers do not necessarily measure gravitational acceleration. Gravimeters on the reference frames may be aimed at remote objects. They must measure the gravitational acceleration of the object at an instantaneous occurrence, like the origin of a solar flare or the strike of a meteoroid on the object's surface. The precise location of the object is unimportant because it is the absolute magnitude of gravitational acceleration that matters. It remains nearly constant in most cases. Even if existing gravimeters and accelerometers do not have the performance required, we assume they will come about someday. In regard to micro space, it is now possible to measure the gravitational acceleration of a two-millimeter-wide spherical object. Improvements may be made in the future. Perhaps other forces, such as the strong force, can be employed rather than gravitational forces to define distances.

The author of this paper is a former space scientist who worked under contracts and grants from several NASA space centers. Much of the research involved so-called strap-down guidance, which made use of accelerometers. Some of his research is cited in Google Scholar under "J.F. Andrus". Also, he was the Manager of the Scientific Sub-Operation of the Marshall Space Center's computer lab during the first moon landings. He was the co-author of a published research paper on the optimal path to the moon.

Initial Conditions

One can assign arbitrary values to the initial three-dimensional position and velocity vectors of the three reference frames. The justification is that we may perform, in our imagination, six operations on the entire universe. These operations consist of two angles of rotation, two angular velocities, one translation, and one translational velocity. Rather than arbitrary choices, one would probably choose conventional values.

The existence of absolute acceleration at all events of the universe provides a rigid spatial structure if initial conditions are specified.

It was shown earlier how the initial position of a non-reference process p may be computed. The initial velocity of p may be calculated by determining the position vector of a second event E_2 of the process and adjusting the initial velocity in order that mathematical integration of the trajectory will yield that position at E_2.

Time

In our model, time is basically different from space: we do not combine it with space as in the general theory of relativity. Each process has its own time, as given by a clock associated with it. A particular reading of the clock is a *feature* of an event of the process. Therefore, a particular *reading* may be considered to be real. If two processes interact (as when two people shake hands), there is no reason to believe that their clocks will agree. It is known that in some situations, the readings may be quite different. Movement through so-called space does affect the time at which processes interact. Each event may be labeled with positions and time coordinates, providing a man-made construction of space and time.

We mention Spinoza's static form of the principle of plenitude in which all genuine possibilities exist in a timeless universe: Why should something exist at one time rather than another? We quote Spinoza (1934): "A thing necessarily exists if no cause or reason can be granted which prevents its existence." We believe that this principle is the very basis of existence. It implies that the universe is timeless. How, then, can time be real?

There are other arguments against time as we conceive it. Is time tenseless? We quote Mellor (1993): "The reality of tense is disapproved by a contradiction inherent in the idea that time flows, i.e., that things, events, and facts really change when these ... change from future to present to past." This argument was given by J.M.E. McTaggart. Of course, there were many philosophers who taught that change is an illusion.

On any process, the position vector r as a function of time t is given by the equation.

$$\mathbf{d}r/\mathbf{dt} = r_0 + (\mathbf{t} - \mathbf{t_0})\, v_0 + \int_{t_0}^{t} \Omega \mathbf{dT}$$

where $\mathbf{d}r/\mathbf{dt}$ is the rate of change of r with respect to time. The symbol Ω stands for total acceleration. The symbols t_0, r_0, and v_0 represent the initial time, position, and velocity, respectively. The symbol S stands for the summation of the area blocks under the antiderivative curve from t_0 to t on the horizontal axis. The antiderivative is the reverse of the rate of change. The arclength at time $\mathbf{t}$ is $\mathbf{s(t)}$. It is the absolute magnitude of $r(\mathbf{t})$. Thus, arclength can be calculated along any trajectory, and r can be given as a function $\mathbf{s}$ (rather than $\mathbf{t}$) as the independent variable. This demonstration does not depend upon the existence of a real space but just upon the formulation of our make-believe space.

How is the direction of time determined in our model? Let us begin with our dog Muffit. Let E_1 and E_2 denote two events in the process P, consisting of events in the life of Muffit. Event E_1 is the event at which Muffit was conceived, and E_2 is any event in Muffit's adult life. We take the forward direction of P to be the one that leads from E_1 to E_2. The forward direction of another process, Q, can be found by determining the events of Q, which affect E_1 and E_2 by means of gravitational forces. This fits in well with memories: events in the early life of Muffit (for example) can be stored (in part) and ordered in the brain of Muffit. Thanks to Muffit the forward direction of time can be determined at the center of many celestial bodies.

Most physicists believe that the direction of time is that of increasing entropy (disorder). This does not seem to contradict Muffit. I'll toss him a bone!

Andrus (1987) discusses what some other authorities have said about the existence of time. We quote them in the following paragraphs:

To use the often-quoted statement of H. Weyl, "The objective world simply is, it does not happen." We are also in close agreement with A. Grünbaum's statement that "...classifying events as past, present or future has no significance apart from the egocentric perspective of a *conscious* conceptualizing organism..."

There is obviously a close connection between our definition of a process and A. Grünbaum's idea that "...processes can be postulated to be aggregates of point-events... . Events simply are or occur...but they do not 'advance' into a preexisting frame called 'time'". We are in complete agreement with these statements. Moreover, we are in complete agreement with Grünbaum when he employs the idea of symmetrical relationships between events. For instance, he states that "Each of the two genidentical events E and E' can therefore be said to sustain a relation of causal connectedness to each other, which is thereby symmetric." The word "genidentical" and the expression "causal connectedness", as employed by Grünbaum, signify a relation of sameness between events containing persistent physical objects.

B.C. van Fraassen, a student of Grünbaum, must have been thinking along these lines when he wrote: "Our conclusion is that it is not necessary to say that there is such a thing as time, but that if we do, the best possible answer to the further question what kind of thing it is, is that it is a logical space." Earlier he defined a logical space as follows: "We characterize the notion of *logical space* by saying that a logical space is a certain mathematical construct used to represent certain conceptual interconnections. ...this does not mean that our logical space must be a mathematical construct isomorphic to the actual temporal structure of events. It is necessary only that the latter can be embedded in the former.

Equivalence of Mass and Energy

Our model assumes Maxwell's field equations (derived empirically). These are discussed by Mazur (2015). They indicate that a pulse of light travels at a constant speed. The speed is known, and it is denoted by the symbol c. Any photon process will experience zero absolute acceleration. Maxwell's equations include electro-magnetic radiation as well as light, which is a component of the electro-magnetic field. These equations imply that the speed of light in a vacuum is constant. Einstein used Maxwell's theory to prove that:

$$E=Mc^2.$$

This is discussed in the Stanford Encyclopedia of Philosophy in 2023.

The twin paradox does not need to be proven. Why should one expect that two processes intersecting at a second event should show the same time? This so-called twin paradox involves twins, one of which leaves Earth, accelerates to a high velocity, and returns to Earth exhibiting a considerably different age than the remaining twin. This effect has been verified using a clock sent into space.

Beginning of the Concept of Space

As living creatures evolved, it was necessary that they be enabled to avoid danger. For example, if a chicken saw a fox, it was to the chicken's advantage to hide. The brightness of the fox was indicative of the magnitude of impending danger. We believe it was light more than gravity that led to the notion of distance.

Brightness and a sense of gravitational force help animals survive in many respects. However, it is not necessary that all of nature correlate in this regard. Quantum entanglement does not correlate because it can involve events that seem to be far away from each other.

General Theory of Relativity

The general theory of relativity was a magnificent accomplishment of Albert Einstein; however, it has come to the point that contradicting the general theory of relativity is tantamount to contradicting Holy Writ. Perhaps concepts, such as those of dark energy and matter, are formulations covering for defects in the general theory of relativity. Moreover, the general theory of relativity does not include all fields of force.

There are skeptics of the general theory of relativity. For example, the Nobel prize winner Laughlin (2005) declared: "It is still controversial and beyond the reach of experiment".

The reverence for Einstein is reminiscent of that of the Greek philosopher Aristotle. Who is one to contradict such a genius? Was he right to believe that earth is the center of the universe? (In a way, he was correct: In order for man to exist, everything must be so as to permit it. This is the anthropic principle.)

To its credit, general relativity predicts exactly the amount of perihelion advance seen in the orbit of planet Mercury. This is in general, relativity space. Here general relativity provides a correction to Newton's law of gravity.

Consider the set S of processes to which our model is able to assign initial conditions, given that the initial conditions of the three reference frames have been assigned and that we have decided upon a definition of mass (see earlier discussion). Set S includes the trajectory of Mercury.

Consequently, the initial conditions of all processes in S are uniquely determined, and we have predicted exactly the amount of Mercury's perihelion advance in our space.

The great preponderance of processes in the universe will not receive labels. The great majority of them do not even fall within our light cone.

General relativity is concerned with the large-scale behavior of the universe, whereas quantum mechanics deals with the small scale. The two theories do not match, but many efforts have been underway to develop a plausible model that includes both theories.

There are other difficulties with general relativity: an adult can meet herself as a child. This conundrum can derive from straightforward solutions to the field equations. We quote Rovelli (2018): "In this way, a continuous trajectory toward the future returns to the originating event. The first to realize this was Kurt Gödel, the great twentieth-century logician ...". Gödel was Einstein's best friend in their old age. It is ironic that he helped undermine Einstein's general relativity.

Although we may consider space and time to be fundamentally different, that does not imply that space does not affect time. Spatial coordinates, in our theory, depend upon acceleration. These, in turn, are features of the events of a process and, consequently, help to determine what are considered to be the events of the process. Therefore, the times assigned are affected, and time is not independent of space.

Kant and the Question of What Exists

Kant (1966) wrote that space and time exist in the imagination of man. He taught that imagination provides the exact geometry of space. The latter idea has been discredited, but we believe that Kant was right to the extent that the imagination provides a strong notion of space.

We also believe that the imagination (or, more precisely, the brain) does the enormous service of placing the world in a form that enables us to cope with the world. The real world may be beyond our comprehension.

Leibniz (1934), too, believed that space was not real. We quote: "As for me, I have more than once stated that I held *space* to be something purely relative, like *time*; space being an order of co-existences as time is an order of successions". Why should it be that events are ordered one way rather than another? (Leibniz was trained in law, but he is better known today for his hobbies. These include mathematics, physics, and philosophy.)

In reference 2, Andrus states that development is not a matter of change over time. It is akin to topographical change; e.g., mountains rise from the foothills to the peaks. As Parmenides taught, change is an illusion.

We believe that three-dimensional absolute acceleration is real, as is brightness and the force of gravity. The imagination of man (including notions of space and time) is assuredly real.

However, any assignment of position coordinates to events is Man-made. Time is not real, but the readings of clocks are real.

We quote Sklar (1974): "Having made these assumptions, are we then committed to the acceptance of a "real" spacetime existing over and above the admittedly existing objects and their admitted spatio-temporal relations to one another … the answer is negative. … Absolute acceleration is a property that a system has or does not have, independently of the existence or state of anything else in the world".

Are logic and mathematics real? We consider that they are real, but they fall within their own category of reality. They are *not* real in that they can interact in conscious events. The category of logic mathematics places considerable limits on what is possible; i.e. the "gods" of mathematics rule over the "gods" of nature.

Are the laws of nature real? We believe – as Leibniz taught – that they are real only in the sense that realities of the universe have features that are limited; it is within their nature that certain happenings cannot occur in later events of their life histories. See Wiener (1951).

We are reminded of my discussion with my cousin from Waynesboro, Virginia. I asked, "Do ghosts exist?" She responded, "No." I then asked, "Do fairies exist?" She again responded, "No." I finally asked, "Do angels exist?" She responded, "Now you are stepping on my toes. I'm a Christian, and it's very convenient to be a Christian around here." Unfortunately, she could not belong to a church because she was "allergic to perfume."

Is God real? A poll has been taken of the members of the National Academy of Science. This group consists of leading American scientists. The 1998 poll showed that 72.2% were atheist, 20.8% agnostic, and 7.0% believed in a personal god.

The Big Crunch

Why should something happen at one time rather than another? In Spinoza's timeless universe, there is no time as a fundamental quality. If time is essentially a man-made thing, we will allow ourselves to choose the so-called forward direction to be either the forward or backward direction.

Let us choose it to be the backward direction as opposed to the conventional forward direction. This enables us to replace the idea of the big bang with that of the *big crunch*. Then gravity in reverse time increases dramatically as matter is drawn toward a central point. Most of the heavy elements exist in the present time. Observation has shown that most of the heavy elements existed within the first million years of forward time. We propose that this was before sufficient mass had accumulated in reverse time to have brought in the big crunch.

The expression "big crunch" has been used in the sense that the universe ultimately collapses.

As explained elsewhere in this paper, we do not need to be concerned with cause and effect because they don't exist.

Evolution seems to contradict our scenario: it is occurring in the wrong direction of time. We will propose a way out of this problem. It brings in the *many worlds theory of quantum mechanics* in which the universe constantly splits into many branches. It is believed that the splitting occurs in both directions of time. The belief is based on the idea that indeterminacy occurs in both directions.

Consider a branch leading to a very high degree of evolutionary development and a small amount of entropy (disorder) in reverse time. Here biological evolution is occurring in the wrong direction, but it is practically negligible compared to the splitting effect. For one thing, order is increased, which is beneficial for life. Both types of evolution make enormous use of mutation, but splitting is much more effective in this regard. Darwin's evolution must work in the back direction because it was not present initially in forward time. It required heavy element chemicals which were not present.

In forward time, nothing existed before the big bang, according to our theory. The crunch was a singularity. As mass increases, time slows. In quantum theory, as mass increases, time disintegrates as a meaningful thing. In the beginning,…well, maybe there was no beginning.

Our theory seems to work well with the idea that we consider the beginning event of the universe to be an event of nearly maximum intelligence. This will occur nearly the same time as the present event in astronomical time. Most of the heavy elements are present.

If we chose the backward direction of time, we happily get younger every day, but (as discussed elsewhere in this paper) we may exist for only an instant.

Intelligence is so valuable to living things that one would expect that the more advanced species would evolve faster than those with little or no intelligence. There is evidence for this.

George Gaylord Simpson believed there was accelerated body size evolution during cold climatic periods in the Cenozoic Era.

In Patterson (1999), a graph is presented which indicates that, from a common ancestor, it took 60 million years for humans to develop, whereas it took 140 million years for crocodiles to come about. We believe this indicates that intelligence brought on accelerated change. In Patterson (1999), another chart indicates that, from a common ancestor, it took 130 million years for placental mammals to develop; but it took 325 million years for crocodiles to come about.

In Martin (1990), based on immunological evidence, the diagram indicates that from a common ancestor, it took 30 million years for old-world monkeys to come about. On the other hand, it took only 20 million years for chimpanzees to develop. Similar results may be derived from the chart based on paleoethological evidence.

We believe that in order for living things to exist, the universe must be of such a nature that some concept of space must exist. We quote Barrow and Tipler (1986): "The basic features of the universe, including such properties as its shape, size, age and laws of change, must be *observed* to be of a type that allows the evolution of observers, for if intelligent life did not evolve in an otherwise possible universe, it is obvious that no one would be asking the reason for the observed shape, size, age and so forth of the universe." This is the anthropic principle.

Music, Art, and Religion

With intelligence came the important developments of speech and language. Appreciation of music, the arts, and religion followed. Religion was needed to hold civilizations together through common values. It kept order by threatening hell, etc. Music and art enhanced cathedrals and temples to further impress the people. In the modern world, government takes over the role of religion in imposing order. Religion has become a danger because of its propagation of irrational beliefs. Similar ideas have been written by others.

Conclusion

Our model of space does not employ tensors or systems of partial differential equations that must be solved as in general relativity. Instead, the trajectories of objects are mainly obtained using simple mathematical integration of acceleration acting upon objects. Our model employs gravitational acceleration to define the distance of an object from an observer. *All* forces, even man-made ones, are included in our model. It offers candidate mechanics to compete with Newton's and Hamilton's.

By reversing the primary direction of time, we theorize that the big bang concept can be replaced with that of the big crunch. A new theory of evolutionary development is proposed in which the many worlds theory is fundamental.

We propose that the brightness of light, more than gravitation, is basic in the development of the intuition of space.

The model employs three reference bodies. These three bodies are required to locate other bodies in space. Moreover, with three reference bodies, it is possible to assign six initial conditions to be assigned arbitrarily or using conventional conditions. This is exactly the number of conditions needed.

Perhaps the anthropic principle is telling us that this is just what is needed for a viable model of space to come about.

We also propose a theory in which the big bang is replaced by the big crunch, working in the opposite direction of time than the conventional one.

My daughter, Leslie, will be typing this monograph, but my wife says it's time to stop the nonsense and take out the garbage.

References

Adams, Steven. 1997. *Relativity: An Introduction to Spacetime Physics*. 111. Philadelphia: Taylor & Francis.

Andrus, Jan Frederick. 1987. "The Time Variable." *The Southern Journal of Philosophy* XXV: 3, 4, 7-12.

Barrow, John D., and Frank T. Tipler. 1980. *The Anthropic Cosmological Principle*. 1-2. New York, NY: Oxford University Press.

Einstein, Albert. 1901. *Relativity*. 155. New York, NY: Crown Publishers.

Jammer, Max. 2000. *Concepts of Mass in Contemporary Physics and Philosophy*. 6. Princeton, NJ: Princeton University Press.

Kant, Immanuel. 1966. *Critique of Pure Reason*. Garden City, NY: Doubleday Anchor.

Laughlin, Robert B. 2009. *A Different Universe*. 119. New York, NY: Basic Books.

Leibniz, Gottfried Wilhelm. 1934. *Philosophical Writings by Leibniz*. 199. New York, NY: E.O. Denton and Company, Inc.

Martin, R. D. 1990. *Primate Origins and Evolution*. 692. Princeton, N.J.: Princeton University Press.

Mazur, Eric. 2015. *Principles of Physics, Part 2*. 806. Boston, MA: Pearson Education, Inc.

Mellor, David Hugh. 1993. "The Unreality of Time". 47. Oxford: Oxford University Press.

Morison, Ian. 1943. *Introduction to Astronomy and Cosmology*. 210-11. The Atrium, Southern Gate: John Wiley and Sons Ltd.

Patterson, Colin. 1999. *Evolution, Second Edition.*. 63, 107. London: The Natural History Museum.

Rovelli, Carlo. 2018. *The Order of Time*. 53. New York, NY: Penguin Random House LLC.

Sklar, Lawrence. 1974. *Space, Time, and Spacetime*. 230-300. Berkley, CA: University of California Press.

Spinoza, Benedict. 1951. *Works of Spinoza*. 52. New York, NY: Dover Publications, Inc. 534

Wiener, Philip P. 1951. *Leibniz Selections*. 534. New York, NY: Charles Scribner's Sons.

www.ingramcontent.com/pod-product-compliance
Lightning Source LLC
Chambersburg PA
CBHW072227290526
45794CB00007B/2914